MW01400323

Extreme Animals

Nature's Strongest Animals

Frankie Stout

PowerKiDS press
New York

For Nicholas Anthony Lazarus, a wonderful nephew

Published in 2008 by The Rosen Publishing Group, Inc.
29 East 21st Street, New York, NY 10010

Copyright © 2008 by The Rosen Publishing Group, Inc.

All rights reserved. No part of this book may be reproduced in any form without permission in writing from the publisher, except by a reviewer.

First Edition

Editors: Jennifer Way and Nicole Pristash
Book Design: Greg Tucker
Photo Researcher: Nicole Pristash

Photo Credits: Cover, pp. 5, 7, 9, 11, 13, 14–15, 17, 21 Shutterstock.com; p. 19 © Michel Gunther/BIOS/Peter Arnold, Inc.

Library of Congress Cataloging-in-Publication Data

Stout, Frankie.
  Nature's strongest animals / Frankie Stout. — 1st ed.
    p. cm. — (Extreme animals)
  Includes index.
  ISBN 978-1-4042-4158-9 (library binding)
  1. Animals—Miscellanea—Juvenile literature. 2. Physiology—Juvenile literature. I. Title.
  QL49S797 2008
  590—dc22
                              2007025675

Manufactured in the United States of America

Contents

Extreme Strength	4
Built for Strength	6
Gentle Giants	8
Big, Strong Bears	10
The Biggest of All	12
African Elephants	14
Small but Strong	16
Power in the Sky	18
Natural Strength	20
Strong Facts	22
Glossary	23
Index	24
Web Sites	24

Extreme Strength

Some animals are known for their power and strength. The stronger an animal is, the better it may be able to live in its **environment**. The world's strongest animals use their powerful body to do all kinds of things that help them **survive**.

An animal does not have to be big to be strong. Some very small animals, like ants and beetles, have a very strong body. Strong animals come in all sizes and shapes. Strong animals also live in all kinds of places, from the air, to the forest, to the seas. From polar bears to gorillas, animals all over the natural world will wow you with their **extreme** strength.

The tiger is the strongest of the world's big cats. This makes tigers the top hunters in their environment.

5

Built for Strength

Strong animals use their powerful body to survive. Their **muscles** are big and their bones are thick. Big muscles and thick, heavy bones can help an animal move things, like trees, out of its way. A strong body can help an animal lift and carry food or fight off enemies.

Muscles are the parts of an animal's body that make an animal move. Muscles are what let an animal get away from danger or get around to look for food. Without muscles, an animal could not move and would die.

Not only muscles, but also size can make an animal strong. A big animal, like the elephant, can put more force on things just because it is so big.

The polar bear lives in a cold Arctic environment. These huge bears are the strongest animals in their environment.

7

Gentle Giants

The mighty gorilla can weigh as much as 600 pounds (272 kg) and stand almost 6 feet (1.8 m) tall. A gorilla's body is thick, with a large chest and hands. Gorillas are the largest **primates**.

Though they can look big and scary, gorillas do not use their strength to kill other animals for food. Most gorillas eat plant leaves and fruit, although sometimes they eat ants and other bugs.

Male gorillas can use their size to scare other gorillas away or to fight. They may stand up on their back legs and puff their chest out. They may beat their chest, too, and make a loud hoot. The biggest, most powerful gorillas are generally the leaders of their groups.

Gorillas are endangered, which means that they are in danger of dying out forever.

Big, Strong Bears

Bears are the largest, most powerful **carnivores** on land. There are different kinds of bears, and all are very strong.

Black bears are about 5 feet (1.5 m) tall and weigh about 400 pounds (180 kg). Polar bears can grow to about 8 feet (2.4 m) and weigh 1,600 pounds (726 kg). Grizzly bears can grow to more than 9 feet (2.7 m) and weigh 1,700 pounds (771 kg). That is 10 times the size of a person!

Bears use their great strength to hunt for food and to keep other animals away. Polar bears eat seals and other **marine mammals**. Black bears and grizzlies eat fish, berries, nuts, roots, and small animals, like young moose and caribou.

The grizzly bear is strong enough to easily move this large log, which would need several people to move it.

The Biggest of All

The biggest land animal of all is the African elephant. Everything about an elephant is big and strong. An elephant's tooth can weigh 3 pounds (1 kg). Big, strong teeth help an elephant chew through 400 pounds (181 kg) of food every day.

Elephants have big, round legs. They need strong legs to carry their weight. Elephants can weigh more than 10,000 pounds (4,500 kg)!

One of the strongest parts of an elephant is its trunk, or nose. Elephants have a long, muscular trunk that lets them lift heavy things, like trees.

The elephant's trunk is strong enough to pull down a tree. The elephant uses its trunk to reach food and water and to move things out of its way.

Elephants can run 25 miles per hour (40 km/h).

African Elephants

Wow!!

The largest elephant weighed 24,000 pounds (10,886 kg) and stood 13 feet (4 m) at the shoulder!

Long-Distance Call

An elephant's call can be heard 5 miles (8 km) away.

Extreme Facts

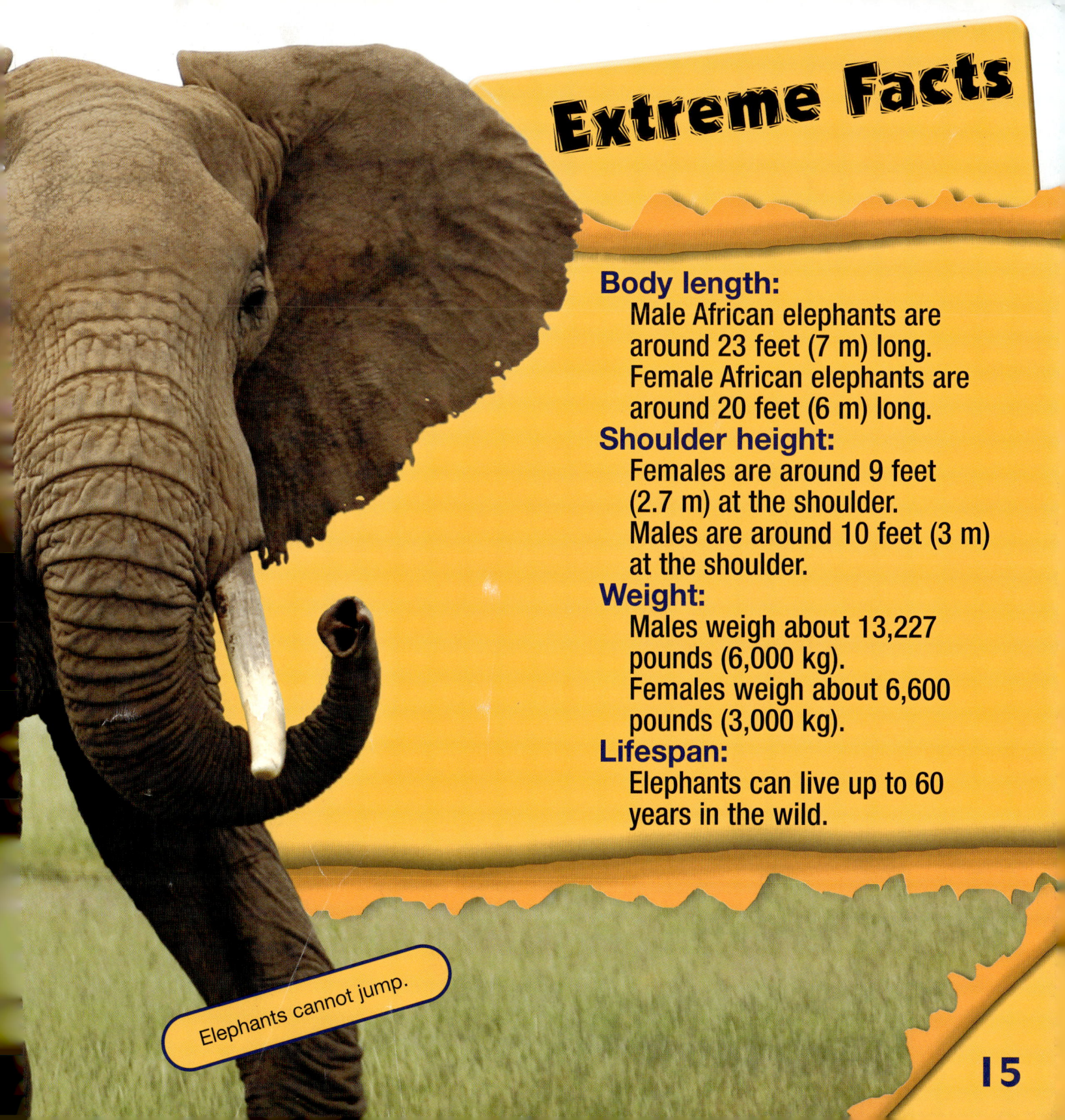

Body length:
Male African elephants are around 23 feet (7 m) long. Female African elephants are around 20 feet (6 m) long.

Shoulder height:
Females are around 9 feet (2.7 m) at the shoulder. Males are around 10 feet (3 m) at the shoulder.

Weight:
Males weigh about 13,227 pounds (6,000 kg). Females weigh about 6,600 pounds (3,000 kg).

Lifespan:
Elephants can live up to 60 years in the wild.

Elephants cannot jump.

Small but Strong

Though they are small, many **insects** are strong for their size. For example, rhinoceros beetles have large **horns** that they use to force other beetles away. They are the strongest insects on Earth. They can carry objects that are 850 times their body weight!

Ants are some of the strongest animals on Earth. An ant can carry up to 50 times its own body weight.

Ants use their strength to carry leaves, **twigs**, and even other bugs back to their nests. Some ants like to eat caterpillars. Army ants carry their larvae, or baby ants, from place to place as they move around looking for food.

If a person could carry an object 850 times his or her weight, like this rhinoceros beetle can, that object would weigh around 64 tons (58 t)!

Power in the Sky

More than any other bird, the eagle is known for its power. Standing almost 3 feet (90 cm) and weighing around 8 pounds (4 kg), the crowned eagle is one of the largest, strongest eagles. The wingspan of a crowned eagle can be up to 7 feet (2 m).

Crowned eagles eat monkeys and other mammals that live in central and southern Africa. The animals that crowned eagles eat can be much larger than the eagles are. Crowned eagles are so strong that they can even catch small antelopes in their talons!

The crowned eagle catches animals that weigh up to around 75 pounds (34 kg). This is about nine times the eagle's weight!

Natural Strength

All kinds of animals use their strength to catch food, to keep other animals away, and to move around. Special muscles and a body built for power can help an animal remain living.

Without their powerful body, ants could not carry food back home. Elephants could not eat without their muscular trunk. Gorillas could not form strong family groups without mighty leaders.

Extreme strength is just one of the many great abilities that animals have. Whether big or small, strong animals use their body to survive.

The ant can move things that are up to 50 times its weight! This ant is carrying a fly's head.

21

Strong Facts

Strongest Animal for Size: The rhinoceros beetle can lift 850 times its own weight.

Strongest Animal of All: An elephant can carry about one-fourth its own weight.

Strongest Person: In 2000, Pyrros Dimas, from Greece, lifted 860 pounds (390 kg). Dimas weighed about 182 pounds (82.5 kg). This means that he lifted almost five times his own weight!

Glossary

carnivores (KAHR-neh-vorz) Animals that eat other animals.

environment (en-VY-ern-ment) All the living things of a place.

extreme (ik-STREEM) Going past the expected or common.

horns (HORNZ) Large branchlike things that grow on the heads of some animals.

insects (IN-sekts) Small animals that often have six legs and wings.

mammals (MA-mulz) Warm-blooded animals that have a backbone and hair, breathe air, and feed milk to their young.

marine (muh-REEN) Having to do with the sea.

muscles (MUH-sulz) Parts of the body that make the body move.

primates (PRY-mayts) The groups of animals that are more advanced than others and includes monkeys, gorillas, and people.

survive (sur-VYV) To live longer than, to stay living.

twigs (TWIGZ) Small branches.

Index

A
air, 4
ants, 4, 16, 20

B
beetle(s), 4, 16, 22
body, 4, 6, 8, 20

C
carnivores, 10

E
environment, 4

F
forest, 4

G
gorilla(s), 4, 8, 20

H
horns, 16

I
insects, 16

M
mammals, 10, 18
muscles, 6, 20

P
places, 4
polar bear(s), 4, 10

power, 4, 18, 20
primates, 8

S
seals, 10
seas, 4
shapes, 4
size(s), 4, 6, 8, 10, 16
strength, 4, 8, 10, 16, 20

T
twigs, 16

W
world, 4

Web Sites

Due to the changing nature of Internet links, PowerKids Press has developed an online list of Web sites related to the subject of this book. This site is updated regularly. Please use this link to access the list: www.powerkidslinks.com/exan/strong/